Canada Goose

I0820282

Red-Tailed Hawk

Rock
Pigeon

Northern Flicker

Wonder is often disguised as a bird.

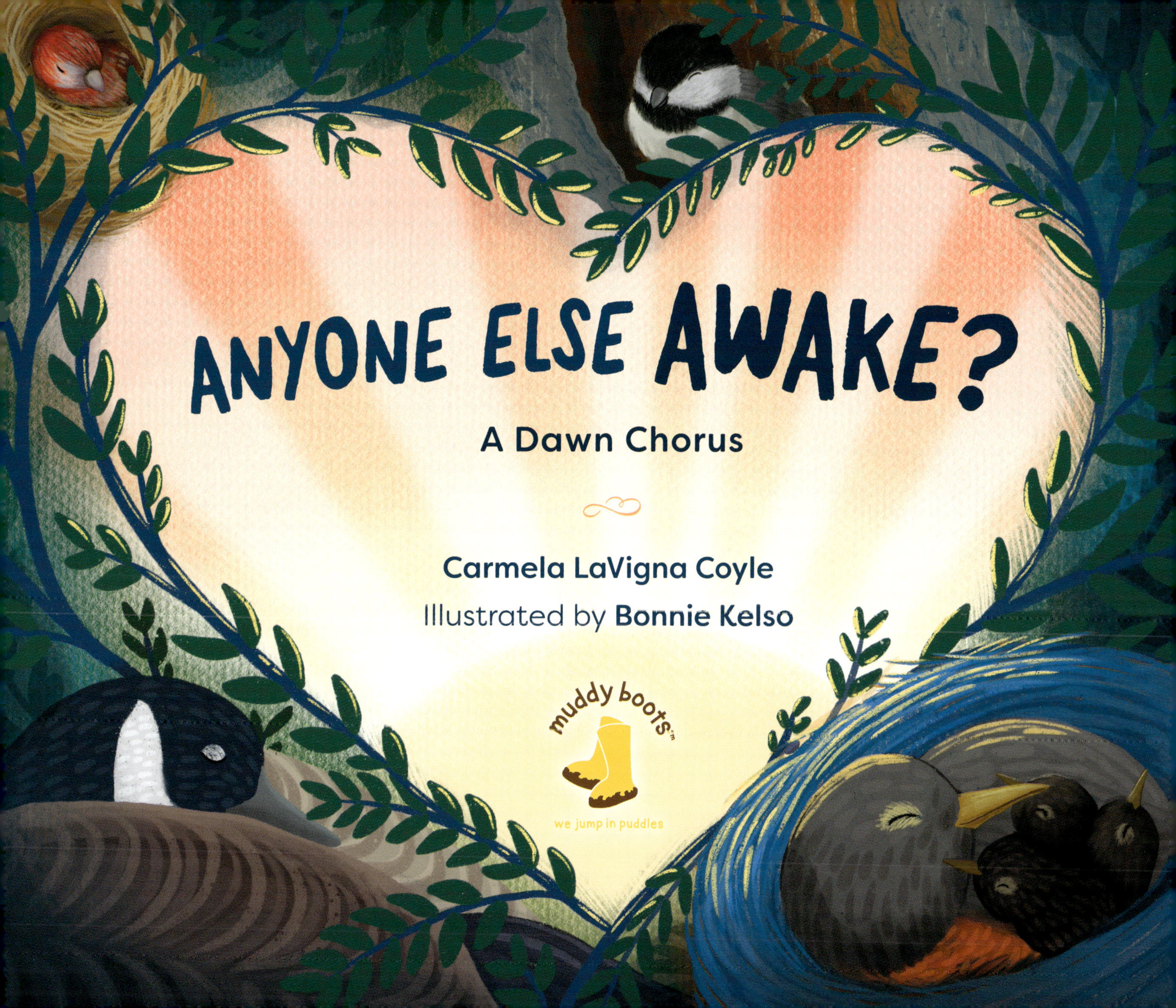

ANYONE ELSE AWAKE?

A Dawn Chorus

Carmela LaVigna Coyle

Illustrated by Bonnie Kelso

muddy boots™

we jump in puddles

An imprint of the Globe Pequot Publishing Group, Inc.
64 South Main Street
Essex, CT 06426
www.muddybootsbooks.com

British Library Cataloguing in Publication Information available

Library of Congress Cataloging-in-Publication Data

Names: Coyle, Carmela LaVigna author | Kelso, Bonnie illustrator
Title: Anyone else awake? : a dawn chorus / Carmela LaVigna Coyle ; Illustrated by Bonnie Kelso.
Description: Essex, CT : Muddy Boots, An imprint of the Globe Pequot Publishing Group, Inc, [2026] | Audience: Ages 4–8 | Audience: Grades K–3 | Summary: "This lyrical book about songbirds describes the unique voices of eleven common species. Helpful back matter provides further information on how and why birds sing the way they do"—Provided by publisher.
Identifiers: LCCN 2025022839 (print) | LCCN 2025022840 (ebook) | ISBN 9781493090570 cloth | ISBN 9781493090884 epub
Subjects: LCSH: Songbirds—Juvenile literature | BISAC: JUVENILE NONFICTION / Animals / Birds | JUVENILE NONFICTION / Animals / General
Classification: LCC QL696.P2 C83 2026 (print) | LCC QL696.P2 (ebook) | DDC 598.8—dc23/eng/20250716
LC record available at https://lccn.loc.gov/2025022839
LC ebook record available at https://lccn.loc.gov/2025022840

Printed in New Delhi, India, August 2025

To all the birdies everywhere.—clvc

For Evelyn Rose.—bk

Is . . .

Is . . .

Is anyone else awake-wake-wake?
warbles a robin.

Is anyone else awake-wake-wake?

I am . . . I am, chants a chickadee.

Is anyone else awake-wake-wake?
I am . . . I am.

Me-too-too . . . hee-hee-hee-hee,
tweets a towhee.

Is anyone else awake-wake-wake?
I am . . . I am.
Me-too-too . . . hee-hee-hee-hee.

SO AM I! SO AM I!
calls a crow.

Is anyone else awake-wake-wake?
I am . . . I am.
Me-too-too . . . hee-hee-hee-hee.
SO AM I! SO AM I!

I think so-o-o-o. I think so-o-o-o,
drones a dove.

Yep-yep-yep-yep-yep, confirms a flicker.

Is anyone else awake-wake-wake?
I am . . . I am.
Me-too-too . . . hee-hee-hee-hee.
SO AM I! SO AM I!
I think so-o-o-o.
I think so-o-o-o.
Yep-yep-yep-yep-yep.

I'm here and I'm so handsome,
look at me-e-e-e-e,
flaunts a finch.

Let's go! Let's go! Let's go!
gab the geese.

Is anyone else awake-wake-wake?
I am . . . I am.
Me-too-too . . . hee-hee-hee-hee.
I think so-o-o-o.
I think so-o-o-o.
SO AM I! SO AM I!
Yep-yep-yep-yep-yep.

I'm here and I'm so handsome,
look at me-e-e-e-e.
Let's go! Let's go! Let's go!
I LOVE you-you-you!
professes a pigeon.
We found—fooood-foood-food,
announces a nuthatch.

WATCH! OUT! WATCH! OUT! WATCH! OUT!
warn the crows.
SEIGE! SEEEEIGE! screech the hawks.

FLEE-ee-ee! FLEE-ee-ee! FLEE-ee-ee!
FLEE-ee-ee! FLEE-ee-ee! FLEE-ee-ee!

Is . . .

Is . . .

Is everyone else all safe-safe-safe?

I am . . . I am.

Me-too-too . . .
hee-hee-hee-hee.

SO AM I! SO AM I!

I think so-o-o.
I think so-o-o-o.

Yep-yep-yep-yep-yep.

I'm here and I'm so handsome,
look at me-e-e-e-e.

Let's go! Let's go! Let's go!
I LOVE you-you-you.
We found—fooood-
foood-food.
Yum-yum-yum.
Yum-yum-yum,
nibble the nestlings.

Is anyone
else awake?

I am.
Me toooo.
SO AM I!

BIRD TALK

WE LOVE VISITING YARDS, PLAYGROUNDS, AND PARKS. If there aren't a lot of native plants, grasses, bushes, and trees in your area, we may need a little help with food, water, shelter, and a place to raise our young, like nesting boxes or bird houses.

Bathing in birdbaths and natural water sources is one of our favorite things. We can put on *quite* a show. We also drink the water in the birdbath, so please keep it clean and fresh, which also helps prevent mosquitoes. *Fun fact: Bees drink water from birdbaths, too!*

During fall and winter, we love to rest on bare tree limbs, dried stalks, and stems of spent plants. See whether you can spot us!

We eat many different foods! Some of us like to snack on bugs, worms, and beetles. Some of us are "ground feeders," which means we like to forage for seeds and such on the ground. Some of us like to nibble on ripe berries found in bushes and trees.

If you have a bird feeder, remember that seeds taste way better when they are fresh and the feeders are clean.

We are wild birds, so we likely won't want people (or animals) to get too close. We hope you'll watch us from a window or at a distance. Binoculars can help!

Did you know that most songbirds migrate to other areas at night in the fall and spring? You can help us find our proper migratory destinations by keeping outside lights off between midnight and dawn. You can learn more from the National Audubon Society (www.audubon.org/lights-out-program). Hint: The best outdoor lights point downward so as not to disturb us in flight when we are passing by.

You won't believe how many different tunes we can sing! You can listen to some of our melodies at www.audubon.org and Cornell Labs' www.allaboutbirds.org.

INTERNATIONAL DAWN CHORUS DAY is observed each year on the first Sunday in May. If you wake up before the sun, you just might hear an outdoor concert like no other!

MORE CHATTER ABOUT BIRDS

- Each line of this story follows the rhythm and beat of real birdsongs and whimsically suggests a translation based on the many reasons birds sing at dawn. Give it a try the next time you hear birds nearby: What do you think they are trying to say?
- Ornithologists study everything about birds. Through their research, they have discovered *some* of the reasons birds sing in the morning: (1) to stake out or claim territory; (2) to attract a mate; (3) to declare health and fitness; (4) to warn of danger; (5) to announce they've located food; (6) to state their location; (7) to admire their own song.
- Male birds, such as the robin in this story, are often the first singers of the day (*"Is anyone else awake?"*). Female robins sing too, but not as early as male birds—especially if the females are incubating a nest.
- Even though birds *sometimes* communicate with other bird species, they choose their mate from birds of the same species.
- When threatened by predators, birds really do have a universal sound of alarm: "*EEEEEEEE-EEEEEEEEE.*" Can you find that part in the book?
- Although birds sing more profusely during spring and early summer mornings, many birds sing and chirp year-round.
- Be sure to listen to birdsong at the end of the day, too. This is when birds begin to settle down in their cozy nests, roosts, and tree cavities.
- A songbird has a unique voice box called a *syrinx* (*SEER-inks*). This allows birds to (sometimes) sing two notes at the same time!
- Of the 10,000 bird species in the world, only half are songbirds. Storks, pelicans, and some types of vultures have no voice at all!
- The birds in this book can be found in various areas of North America. The next time you go outside, listen for their songs. Can you find the names of the birds who sing near you?

THE THREE PHASES OF DAWN

Astronomical dawn begins when the sun is 18 degrees below the horizon. At this time, it is quite dark.

Nautical dawn finds the sun at 12 degrees below the horizon. At this time, there is just enough light to distinguish some objects from others but not all.

Civil dawn begins when the sun has reached 6 degrees below the horizon and objects can be clearly seen.

Sunrise officially occurs when the sun first peeks out from the horizon.

A FEW CHIRPS FROM THE AUTHOR

I have been observing bird songs, peeps, chirps, tweeps, and chatter ever since I was a young girl. Back then, I didn't realize that the extra boisterous birdsong in springtime had an official name: *The Dawn Chorus*. I did, however, notice birds began singing before the sun rose and often from the tallest tree.

My fascination with birdsong continued as I grew. I would set my alarm and head out on a run to observe the birds and dawn/twilight's three phases. Soon, I began writing outside on a grassy slope next to our house. I journaled about the vibrant colors, the fresh air, the trees, the ever-changing sky, the first sunbeams, and, of course, the birds. To me, it seemed as if the birds were singing "Good morning!" But, of course, I learned it was so much more than that! To this day, I wake before dawn. And many of those mornings I end up outside, with the birds.

Black-Capped
Chickadee
Spotted
Towhee
American Crow
Eurasian Dove
American
Robin
White-Breasted
Nuthatch
House Finch
Nuthatch
Hatchlings